Gem Care

Text and Photographs
Fred Ward

Editing
Charlotte Ward

Table of Contents

4-5	Introduction to Gem Care
6	Amber
7	Beryl (except Emerald) (aquamarine, goshenite, green beryl heliodor, morganite, red beryl)
8-9	Chalcedony & Quartz (agate, amethyst, aventurine, carnelian, cat's eye, chrysoprase, citrine, jasper, onyx, petrified wood, rock crystal, rose quartz, smoky quartz, tiger's eye)
10	Coral
11	Cubic Zirconia
12	Diamond
13	Emerald
14	Garnet (almandite, andradite, demantoid, grossularite, hessonite, hydrogrossular, malaia, pyrope, rhodolite, spessartite, tsavorite)
15	Ivory
16	Jade (nephrite, jadeite)
17	Lapis Lazuli
18	Malachite
19	Metals (gold, platinum, silver)
20-21	Miscellaneous Gems (alexandrite, apatite, azurite, cat's-eye chrysoberyl, chrome diopside, danburite, hematite, iolite, kunzite, marcasite & pyrite, moldavite, moonstone, rhodonite, St. Helens glass, zircon)
22	Opal
23	Pearls & Cameos
24	Peridot
25	Ruby & Sapphire
26	Spinel
27	Tanzanite
28	Topaz (blue, colorless (or white), golden, imperial, pink)
29	Tourmaline (green, indicolite, parti-color, rubellite, watermelon)
30	Turquoise
31	Mohs Hardness Scale
	Index

Introduction

To help you understand and wisely care for your jewelry, I have prepared this unique comprehensive directory as a companion publication to my GEM BOOK SERIES. The gems are indexed and listed alphabetically. Across the top of each page is a bar color-coded for specific gems. Each page begins with a brief description of the gem, continues with a paragraph on care, and ends with a paragraph on cleaning procedures. At the bottom of each page is a synopsis of the gem's most important properties and care features.

You may wonder why gem and jewelry care is important. After all, if gems have already lasted millions of years, why do they need special care? Once mined and worn, they are exposed to hardships and chemicals they never experienced underground. Softer and more vulnerable gems require extra attention. And all gems need to be kept clean to display their beauty fully, which is, after all, the main reason people love gems.

There is little care and cleaning information in print, and unfortunately, much of what is available is incorrect. You can safeguard your gems and jewelry if you follow some important precautions. A few minutes of care will reward you with years of enjoyment.

As you go through this book, you can learn a great deal about gemstones by paying attention to one quality—hardness—not the only determinant of quality in a gem, but extremely desirable. To measure hardness, we use the Mohs Scale, a gem-trade standard. Named for Friedrich Mohs, who conceived it in the early 1800s, the scale is structured so that material rated at each higher number can scratch everything with lower numbers. After determining that diamonds are the hardest substance, Mohs placed them at 10. Rubies and sapphires are Mohs 9, topaz is Mohs 8, quartz is Mohs 7, and talc, the softest, begins the scale at 1.

Most of the objects you and your jewelry touch during the day are either quartz-based or near the hardness of quartz. If your gems are harder than quartz, they will not be scratched; if they are softer than Mohs 7, they can be. Rings and bracelets are most vulnerable, then pendants, pins, and earrings. You may be surprised that the gold, silver, and platinum in your jewelry are only Mohs $2^{1}/_{2}$ - $4^{1}/_{3}$, which means that almost every single gem in your collection is harder than the metal that holds it. A few years in contact with gems leaves gold, silver, and even platinum covered with scratches. Unless you keep your precious metals away from other metals, rocks and dirt, walls, stoves, and counters, expect to do some repolishing. Hardness is important to you as a jewelry owner because more damage will likely occur to your gems and precious metals from contacting harder gems than from accidents.

To Gem Care

Take off your jewelry before you do housework or gardening. Be especially careful with rings. Never remove jewelry by pulling directly on any of its gemstones. Check for loose gems before each wearing by gently shaking the piece or by tapping it with your finger near your ear. Have all loose stones tightened before wearing the jewelry. Check clasps and fasteners often. Restring necklaces regularly, at least every two years, and annually with heavy use. Knot beads and pearls to protect them. Do not store your valuables in heaps. Wrap each piece of jewelry in velvet, paper, or silk. Many people tumble jewelry into plastic bags or bedroom drawers, letting the pieces they have just worn drop onto other jewelry. Diamonds, rubies, and sapphires will scratch or abrade every other thing they touch.

Scratching can also occur from improper cleaning. Do not overclean. Most gems do not need cleaning after every use, except pearls and other porous materials which may be harmed by residues of perspiration, perfume, cosmetics, and spray. Choose chemicals, brushes, and techniques for the softest and most vulnerable component. Jewelry other than all-metal pieces has components of different hardnesses. Toothpaste is often incorrectly used as a jewelry cleaner, but some toothpastes contain abrasives near the hardness of quartz. Those can damage your softer gemstones and sandpaper the precious metal in your jewelry to a matte finish.

Use strong commercial products only when you are sure they will not damage your jewelry. Although most crystalline gems sparkle after applications of commercial cleaners, porous gems, such as pearls, malachite, and turquoise, may be discolored or damaged by cleaning fluids they absorb. For instance, if you have a diamond and pearl ring, use only fluids and techniques that are safe for the pearl. Silver polish works fine for sterling, but it can ruin porous gems set in silver. Ammonia is a particular threat; only harder, nonporous crystalline gems can withstand it. When in doubt, do not use chemicals of any sort.

Regular use of warm water, a very small amount of plain bar soap (not a liquid or powdered detergent), and an old, soft toothbrush is an unbeatable cleaning combination. Rinse to remove the soap, and dry. Warm soapy water is the most economical and simplest solution for most gems, except for diamonds. Diamonds are magnets for soap and grease, which make them look dull. To reduce grimy buildup, remove diamond jewelry before washing your hands. Ultrasonic cleaners, vibrating at hundreds or thousands of times a second, work safely on diamonds and a few other gems. Check the gem list here before placing your jewelry into an ultrasonic cleaner or steamer.

Gems are a gift from the earth and the sea. Care and regular cleaning will keep your jewelry safe and beautiful, assuring it a long life.

Amber

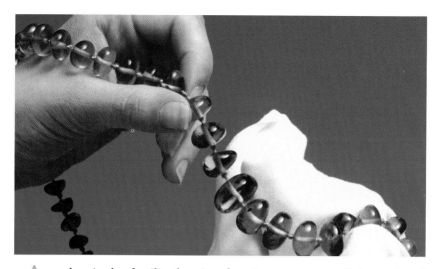

Amber is the fossilized resin of ancient pine trees. It has enjoyed unprecedented popularity and accompanying price rises as a result of the attention it gained from *Jurassic Park*, the book and the motion picture. The story's premise is that dinosaurs can be grown today from DNA found in the stomachs of insects trapped inside amber for millions of years. After the movie was released, wholesale prices soared for amber pieces containing insects. Complete insects in amber are natural and relatively rare. Common internal features seen in amber as disk-shaped sun spangles, said to add character, are almost always intentional fractures created by heating amber after it is mined.

Amber can occur wherever pines grow. Principal commercial sources are the Baltic coasts of Germany and Poland, the Dominican Republic, and new finds in Colombia. Because it is derived from living material, amber is an "organic" gem, along with pearls, coral, jet, and ivory. It requires gentler treatment than crystalline gems. If you have an amber ring, be careful, because it scratches as easily as plastic. In fact, imitation amber usually is plastic. To test for amber, heat a pan of water and stir in table salt until no more will dissolve. Allow the salty water to cool to room temperature. Amber, which is very light, floats in saturated salt water, whereas most imitations sink.

Clean amber only in warm soapy water, rinse in water, and pat dry with a soft cloth. Do not soak. Use no commercial cleaners, chemicals, or brushes. Remember, amber is soft and has to be treated carefully.

Ultrasonic: do not use
Steamer: do not use
Warm soapy water OK; no brushes
Hardness: 2 - $2\frac{1}{2}$
Heat: even low heat will melt amber
Toughness: poor

Solvents can dissolve amber
Keep away from all chemicals
Use no abrasives
May darken with age
Avoid perfume and cosmetics
No mechanical cleaning

Beryl (except Emerald)

Most gems in the beryl family are hardy enough for easy care. Aquamarines (above) are usually ultrasonic-safe. Wash other beryls, including golden beryls (right), to keep them clean.

Beryls make up a family consisting of seven gems, including emeralds, which we deal with separately on page 13. Most gemstones occur in families, but some, like diamonds and spinel, stand alone as single examples of a specific crystal form. Typically, trace elements define separate identities within a gem family, such as color variations, as well as other important physical differences. Members of the beryl family, all of which are beryllium crystals, share basic chemistry and many characteristics with emeralds. Most beryls grow into large, relatively clean crystals. But the trace elements chromium and vanadium, which produce the green in emeralds, cause more serious inclusions and fracturing than the trace elements that color other beryls.

Green beryl serves double-duty. Although sometimes set as a gem, its primary use is as a raw material for creating aquamarine. Heating green beryl transforms it into a clear sea-blue. Often flawless, aquas are sometimes huge. After emeralds, aquamarines are the best known beryls. Lesser-known are golden or orange heliodor; morganite, a pink gem named for J.P. Morgan; red beryl (or bixbite), from Utah; and colorless goshenite.

All other beryls are usually tougher than emeralds and can be cleaned differently from their green sisters. Beryls that are not heavily included or fractured are usually safe in ultrasonic cleaners. If you have concerns about safety, gently clean beryls with warm soapy water, rub with a soft cloth or a soft toothbrush, rinse in warm water, and lay on a towel to dry. Do not scrub. Excessive cleaning can be harder on gems than no cleaning at all.

Ultrasonic: risky with included stones; usually safe with aquamarine
Steamer: do not use on included gems   Resistant to most chemicals
Warm soapy water, soft brush OK   Light: orange may fade to pink
Hardness : $7\frac{1}{2}$ - 8   Toughness: good
Heat: may cause fracturing, breakage, fading, or color changes

Chalcedony & Quartz

The quartz family is filled with attractive color varieties we use as gems. Amethyst (above, in ultrasonic cleaner) is prized for its rich purple shades. Carnelian and onyx (below), typically unfaceted, are best cleaned with a damp cloth. Agate cameo (left) can withstand brisk washing.

Chalcedony and quartz comprise the world's largest mineral group. The difference between quartz and chalcedony is how the crystals form. Quartz grows as single crystals. Chalcedony, the cryptocrystalline form, is a collection of multiple fine-grained quartz crystals. Quartz is best known in its colorless, smoky, or rose forms, but, like chalcedony, it can display a surprising color spectrum.

Usually transparent to translucent, carnelian is red to orange chalcedony. Sard is translucent reddish-brown. Chrysoprase, the light green chalcedony, is often confused with jadeite. Bloodstone, or heliotrope, is recognized by its brownish red spots over a dark green body color. The agates—landscape, moss, iris, and fire—are distinguished by colorfully curved or angular bands that produce the gems' character. Fire agate has an iridescence that reminds some owners of opal. Agates are often used for jewelry, cameos, and carvings. Jasper, of almost any color, tends to be more opaque than other chalcedonies.

Occasionally, as logs deteriorate over vast periods of time, chalcedony replaces the wood fiber, molecule by molecule. The result, "petrified wood," is a colorful agatized replica used for jewelry and tabletops. There is an almost colorless, milky chalcedony that is sometimes confused with moonstone. Onyx is actually an agate-like chalcedony composed of multicolored bands. Virtually all onyx sold today is chalcedony dyed opaque black.

Citrine (right) can be washed under a faucet or wiped clean. Petrified wood (above) or aventurine (left) can be washed or wiped with a damp cloth.

    Crystalline quartz refers to the single crystal form, including such gems as amethyst (purple to violet), rose quartz (pink), smoky quartz (dark brown, gray to black), citrine (yellow to orange), aventurine (green with reflective inclusions of hematite or mica), cat's eye (greenish-yellow to brownish-green, with chatoyancy), tiger's eye (brownish-yellow to brown, with chatoyancy), and colorless rock crystal.

    Generally, quartz is a hard, tough material that can withstand considerable use. Still, because it is softer than the big four gemstones, diamonds, emeralds, rubies, and sapphires, always store quartz jewelry separately to avoid scratches from your other gems. Individual gems in the quartz-chalcedony group have different care requirements, usually associated with light and heat. Some amethyst may fade very slowly in the sun. In contrast, most quartz and other gems do not fade; their colors are a result of light being filtered by the crystals. Avoid a sudden temperature change, which, under certain conditions, fractures quartz. High heat encountered in setting stones may affect the colors of amethyst and citrine.

    The best care for quartz and chalcedony is washing in warm soapy water, which is safe for both the gems and precious metals that hold them. Rinse with clear water, wipe dry, then air dry on a towel. Avoid acids.

Ultrasonic: usually safe   Toughness: good
Steamer: risky   Affected by some acids
Hardness: 7   Warm soapy water OK
Heat: gentle heat lightens amethyst; strong heat can turn amethyst colorless, yellow, or green and can turn citrine colorless
Light: amethyst and rose quartz may fade very slowly in bright light

Coral

Coral is harvested from reefs that grow in tropical seas around the world. The safest care for coral, which is soft, is a gentle wipe with a damp cloth. Avoid all chemicals, and rinse with water when needed.

Coral, like amber, jet, pearls, and ivory, is an "organic" gem, one that derives from living material. In this case, coral is the calcium carbonate "house" that a coral polyp builds in much the way that oysters secrete their protective shell homes. The astounding difference is that coral stalks or trees are the collective homes constructed to exacting architecture by thousands, millions, or billions of communal polyps over decades or centuries. Divers harvest those red, pink, orange, white, purple, and black branches to produce gems.

Two of coral's characteristics determine its care: softness and porosity. As one of the softest of gems, coral scratches and abrades easily. Furthermore, bright light has a tendency to darken it, and it may also be damaged by heat and flames. Store coral in even temperature and humidity. Keep coral jewelry in cloth bags, separated from other gems.

Coral is a porous alkaline material. Acids and solvents soften, swell, melt, and dissolve it. Therefore, keep coral away from chlorinated swimming pools, alcohol, ammonia, ether, turpentine, nail polish remover, and other chemicals. When cleaning your coral jewelry, avoid brushes, abrasives, and all contact with harder materials. Gently wipe your coral with a soft, clean, cool, damp cloth.

Ultrasonic: do not use
Steamer: do not use
Soft clean cool damp cloth; no brushes
Hardness: $3\frac{1}{2}$ - 4
Heat: even low heat can damage coral
Toughness: fair

Acids dissolve coral
Avoid all chemicals
Dry carefully; do not soak
May darken with age
Avoid perfume and cosmetics
Soft, delicate, and light-sensitive

Cubic Zirconia

Cubic zirconia, which are promoted heavily by television shopping channels, are "created" or "laboratory grown" simulants, not natural gems. Of all the materials that have been used for centuries to look like diamonds, CZs are the best yet. Adding trace elements produces every imaginable sparkling color. At Mohs $8\frac{1}{2}$ they are harder than other gems except rubies, sapphires, diamonds, and chrysoberyl. And their unusually high refractive index of 2.15 is high enough to rival diamond's 2.41, giving CZs more glitter than most genuine natural gemstones.

Remember to store them away from harder gems to keep the CZs from being scratched, and away from softer gems so your CZs will not scratch other stones or themselves. Earlier CZs, made more than a decade ago, had a tendency to either fade a bit in sunlight or brown slightly with age. Improved manufacturing techniques assure that today's CZs are stable to light and time. They are now even used for grading diamond color.

Cleaning CZs is at least as easy and in one case even easier than cleaning diamonds. Because they do not have diamonds' cleavage planes, CZs can be cleaned by physical or mechanical means. They are resistant to chemicals, to normal jeweler's torch temperatures, to ultrasonic vibrations, and usually to the effects of bright light. You can clean CZs almost any way you like. Alcohol, vodka, ultrasonic cleaners (above photo), steamers, running water, warm soapy water, brushes, or cloths all work. CZs are very stable materials, and are inexpensive alternatives to diamonds. They do not attract grease the way diamonds do, thus they are easier to keep clean. A little regular cleaning will guarantee years of satisfaction and beauty.

Ultrasonic: safe
Steamer: safe
Hardness: $8\frac{1}{2}$
Heat: sensitive only at very high temperatures
Toughness: good, unless crystal is highly strained

No reaction to chemicals
Warm soapy water OK
Light: stable

Diamond

Diamonds are usually durable enough to withstand ultrasonic cleaning (above). Gems with months of accumulated grease (top left) look like new after just a few minutes' treatment (left).

Diamonds are synonymous with brilliance and durability. Everyone knows two things about diamonds: they are the hardest substances on earth, and they are forever. You might then assume that diamonds are indestructible, which is definitely not true. Diamonds are hard enough to scratch everything, including themselves. But they certainly can break. Because diamonds grow with cleavage planes, a sharp blow on a vulnerable spot can send fragments flying. Less well known is that diamonds are already 600 million to more than three billion years old and that they come in a rainbow of colors, including blue, pink, yellow, orange, and green.

To keep diamonds sparkling and unscratched, separate them from other diamonds. Avoid heat and chemicals on diamonds with fractures that have been filled with sealants, an increasingly common practice.

Diamonds are magnets for grease. To assure brilliant gems, remove the greasy residue that builds from skin oils, soap, and airborne particles. You can use ultrasonic cleaners (unless the diamond has fractures or is fracture-filled), alcohol, ammonia-based cleaners, or other mild solvents. The easiest cleaning procedure for a dazzling improvement is to soak diamonds in vodka for a few minutes, rub with a soft toothbrush, and rinse in warm water.

Ultrasonic: safe unless gem has fractures No reaction to chemicals
Steamer: same as above Hardest of all materials
Warm soapy water and brush OK Grease-cutting detergent OK
Hardness: 10 No direct jeweler's torch
Heat: diamonds begin to vaporize in oxygen-rich atmosphere at 690-875°C.
Toughness: tough in cleavage directions; exceptional in other directions

Emerald

Emeralds are harder than jade, quartz, and some steel. But they can be brittle. The general assumption is that these gems are fragile and easily broken. Actually, few gemstones exceed emeralds in hardness. Emeralds almost always have inclusions, internal growth features that give them character but make them more vulnerable to breaking. Emeralds should be set into jewelry carefully, preferably in bezels for protection. Remove your jewelry before engaging in active sports or heavy work. Let caution and common sense be your guide.

One aspect of emerald processing requires attention. To obscure inclusions, virtually all natural emeralds are oiled during their cutting and marketing phases. A final oiling takes place immediately after cutting, before dealers, manufacturers, and jewelers receive the gems. Long exposure to desert air, the sun, or dry safe deposit boxes evaporates the oils in emeralds. Because strong solvents and even alcohol have a tendency to dissolve oils, do nothing that will remove any residual oil, and have your emerald re-oiled by a jeweler every few years. A main cause of emerald damage is hitting rings against counters. Be especially careful with your emerald jewelry.

Clean emeralds gently with a soft damp cloth, warm water, and a soft brush. Synthetic or created emeralds, usually less included than naturals, are not oiled. Synthetics may be cleaned the same ways as natural emeralds.

Ultrasonic: do not use
Steamer: do not use
Warm soapy water, soft brush OK
Hardness: $7\frac{1}{2}$ - 8
Heat: may cause fracturing or breakage
Toughness: poor to good

Solvents may remove oils
Resistant to most chemicals
Do not store in dry conditions
No direct jeweler's torch
Light: stable
No mechanical cleaning

Garnet

Rivaling emeralds, brilliant green tsavorites from Kenya and Tanzania (top left) are the rarest and most valuable garnets. The design of the red pyrope ring (bottom left) reminds us of the nineteenth century, when garnets enjoyed great popularity. The most affordable and available garnets are purplish-red rhodolites (above, in ultrasonic cleaner).

Garnets are one of the genuine happy surprises in the gem world. Instead of occuring in only one color, garnets form one of the largest and most varied gem families. Most of the garnets sold today are plentiful and inexpensive rhodolites. These purplish-red or reddish-purple gems are commonly seen as beads or in starter jewelry for teenagers. Larger pyropes, up to two carats, can be delightfully red. They are beautiful reminders of the Victorian and Edwardian Ages.

Almandite garnets are reddish-orange to red. Andradites may be green, yellow, brown, or black. The best known andradite is the stunning green demantoid garnet. But the most valuable garnet of all is tsavorite, the spectacular deep emerald-green grossularite garnet. Yellow-orange hessonite is the other grossularite. Hydrogrossular garnets are usually pale green with dark spots. East Africa produces a wonderful pinkish or reddish-orange malaia garnet. The reddest garnets, pyropes, are most often from Czechoslavakia, Australia, and South Africa. Yellowish to reddish-orange spessartite, which is seldom seen, completes the garnet family.

The safest cleaning of garnets is with warm soapy water followed by a water rinse. Alcohol, vodka, or a commercial cleaner may help loosen tough deposits. Use a soft brush on stubborn areas.

Ultrasonic: usually safe except for andradites with liquid inclusions
Steamer: do not use   Resistant to most chemicals
Warm soapy water OK   Toughness: fair to good
Hardness: 7 - 7½, except andradite at 6½ - 7 and hydrogrossular at 7
Heat: abrupt changes may cause fracturing; pyrope fuses with high heat

Ivory

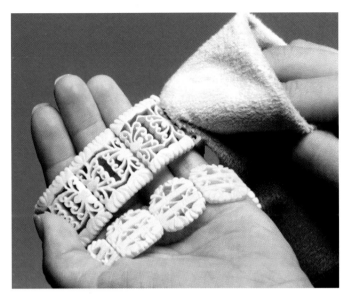

Soft and porous, natural ivory (left) requires regular care and cleaning. Fossil walrus ivory (above) is as hard as quartz and wears as well.

Ivory, an "organic" gem, shares many characteristics with coral and pearls, which derive from living creatures. Although ivory is generally assumed to come from elephants, it also can originate from other large mammals, including whales, walruses, hippos, or even wart hogs. Elephants are now protected because they were hunted until their survival was threatened. Trade in new elephant ivory is prohibited in the United States and much of the rest of the world. However, there is a considerable amount of legal ivory in use and available, as well as new carvings and beads made from the teeth and tusks of other animals. Fossilized ivory from extinct animals, such as mastodons and mammoths, exhibits complex multicolored patterns.

No matter the source, follow certain procedures to protect ivory. Avoid rough wear, heat, and solvents. Ivory abrades easily in rings. It yellows with age and long exposure to bright light. Because ivory is alkaline, many acids shrink, discolor, or dissolve it. Avoid cosmetics, hair spray, and perfume.

Use a damp cloth to clean natural ivory, or at most, a mild soap or detergent with warm water. Do not soak natural ivory; pat it dry, then let it air-dry on a towel after cleaning. Treat scrimshaw more carefully than other ivory; avoid all liquids, which might dissolve the color in markings. Because fossilized ivory is composed of chalcedony, refer to pages 8-9 for its care.

Ultrasonic: do not use
Steamer: do not use
Warm soapy water OK
Hardness: $2^{1}/_{4}$ - $2^{3}/_{4}$
Heat: causes shrinkage and discoloration
Attacked by many chemicals
Toughness: fair
Soft, delicate, and porous
Light: yellows with age
Avoid perfume and cosmetics
Wipe clean with damp cloth; do not soak or bleach; store in dry conditions

Jade

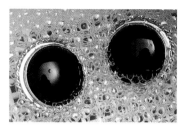

Hard, tough materials, both types of jade respond well to almost all cleaning methods. The white nephrite talisman (left) is perfectly suited to a faucet-water bath. Black jadeite earrings (above center) are safe in an ultrasonic cleaner. And a lavender jadeite pendant from Burma (above right) is spotless after a good wash and rinse under running water.

Jade, the "Stone of Heaven," is revered by a quarter of the world's people as the most valuable and worthy of all the earth's treasures, whereas the rest of the world considers it little more than a rock. A bewilderment to many people is that "jade" is the only gem name that refers to two distinctly different materials: jadeite and nephrite. Today's jade buyers find expensive jewelry is typically jadeite, whereas less expensive beads and carvings are more often nephrite. Another complexity is that China routinely sells approximately thirty other types of rock as "jade." When shopping for jade, an educated buyer has an advantage.

Nephrite, enjoying a 5000-year tradition, is the honored jade. Originally found in what is now far western China, today's nephrite supply is mined mainly in British Columbia. Some is also available from Siberia. China has always valued white nephrite highest, but most nephrite mined today is dark green, although it occurs in black and blue-green as well. Nephrite, the toughest material in nature, was used for tools and weapons for millennia; still, thin pieces may break if dropped. Although the Central American Maya employed jadeite for centuries, it was introduced from Burma to China and the West in the mid-1700s. An instant court favorite in Beijing, jadeite has since completely overshadowed nephrite for jewelry use.

Even though the two jades are different stones, their care and cleaning are similar. Because they are tough and medium hard, you can wash, wipe, and gently rub both types. Jades, like diamonds, require little care other than cleaning to maintain their beauty.

Ultrasonic: safe
Steamer: safe
Slightly affected by warm acids
Warm soapy water OK
Hardness: jadeite $6\frac{1}{2}$ - 7; nephrite 6 - $6\frac{1}{2}$, which is harder than some steel
Heat: jadeite can fuse under jeweler's torch; nephrite is very heat resistant
Toughness: jadeite is exceptional; nephrite is superlative

Lapis Lazuli

The rich blue of lapis lazuli, often flecked with pyrite, makes a dramatic accent for other gems and clothes. Long treasured as a carving and inlay material, lapis is now frequently seen in contemporary jewelry. Wipe clean after every use, and use warm water (with or without soap) when needed.

Lapis lazuli is a richly colored royal blue rock widely used in jewelry. Composed mainly of lazurite, calcite, and pyrite, it is recognized by its blue background flecked with white and veined or flecked with gold-colored pyrite. Lapis is often cut into necklace beads or domed cabochons for rings and earrings. The principal source for jewelry-grade lapis is Afghanistan, although some material is mined in Chile and Russia.

Lapis lazuli and some mineral aggregates are similar to organics in that they are softer and less durable than the harder crystal gems, diamonds, emeralds, rubies, and sapphires. Airborne dust, walls, steel, and sinks tend to have a hardness of Mohs 7 - 7$\frac{1}{2}$. Any gem softer than abrasive quartz, silicon, sand, and most metals is at risk when it comes into contact with the harder substances. At Mohs 5 - 6, lapis should be worn with care. It is a fine choice for necklaces, earrings, and pins, but be especially gentle with lapis bracelets and rings.

The safest cleaning procedure for lapis lazuli is to wash it with warm soapy water and rinse in warm water. Do not scrub. Wipe with a soft cloth, and let it continue drying on a towel. Because lapis is porous, it is best not to soak it or leave it wet. If you are cleaning a lapis necklace, wash it first, pat it dry, then stretch it out fully on a towel for air-drying. This procedure lets the string dry straight.

Ultrasonic: considered risky
Steamer: do not use
Warm soapy water OK
Toughness: fair
Light: stable
Hardness: 5 - 6
Heat: can sometimes darken and improve light material; high heat is risky
Chemicals should be avoided; do not soak or bleach; decomposes in acids

Malachite

Malachite is a bright, cool green, beautifully patterned mineral found in several African countries, Europe, Russia, and Australia. Long used as a decorative stone because of its intricate repetitive circular patterning, it is also popular for cabochons, beads, carvings, and inlays. Malachite became a gem, not for its durability or rarity, but for its unique look. Graceful alternating swirls of color and angular banding give the appearance of an exquisitely painted contour map.

Because malachite is relatively delicate, there are certain precautions you should take to assure years of satisfactory use. Most rocks, minerals, and aggregates are softer than crystalline gems. Lack of durability does not preclude use as gems, but it does require that you be more careful than you might be with diamonds, rubies, or sapphires. If softer gems are protectively set, they can be used in necklaces, pins, and bracelets. Extra care is required with rings. You must be gentle with malachite.

The safest procedure for cleaning is to wash malachite in cool water with a small amount of soft soap. You might also use a very soft brush. Do not soak, bleach, or apply any chemicals. After washing, dry your brooch or bracelet with a clean, soft cloth, and let it rest on a towel to air-dry. When cleaning a malachite necklace, wash first, then carefully pat the beads dry. Stretch the necklace on a towel for air drying. This will let the string dry straight. Cotton strings, which hold more dirt and oil than silk, should be replaced at least every two years.

Ultrasonic: do not use
Steamer: do not use
Cool soapy water OK
Toughness: poor
Light: stable
Hardness: $3\frac{1}{2} - 4$
Heat: avoid heat and sudden temperature changes
Chemicals should be avoided; malachite is attacked by acids; use no solvents

Metals

Metals have been successfully used to hold gems in place for the several thousand years that people have been making and wearing jewelry. While potential competitors (wood, leather, copper, iron, steel, teeth, horns, rocks) came and went, the three early favorites, gold, silver, and platinum, remain today as the carriers of choice. Because they are valuable, malleable, ductile, rare, and beautiful, they are in the broadest sense called *noble metals*. Although silver tarnishes, gold and platinum resist oxidation and corrosion.

Care and cleaning are dependent on what, if anything, is attached to the metal. Any cleaning solution has to be safe for the most vulnerable part of the jewelry. Ultrasonic cleaners are generally safe for noble metals, but if the jewelry includes gems, follow the recommendations for individual stones. Although we think of metals as hard, jewelry metals, only Mohs $2^{1}/_{2}$ - $4^{1}/_{3}$, are softer than almost all gems. Warm soapy water can be safely used to clean all three metals. Alcohol (or vodka) dissolves tough dirt which can then be brushed away with a soft toothbrush. Do not use toothpaste or any other abrasive, because it will scratch jewelry metals. A 50-percent mixture of Mr. Clean or Wisk with water works well.

Gold: the higher the gold percentage (up to pure gold, 24 kt., at Mohs $2^{1}/_{2}$), the softer the metal. Gold is alloyed with copper and other metals to harden it and to reduce concentrations to 14 and 18 kt. (Mohs 3). Use no ammonia, which turns solder joints dark, or mercury, which turns gold white.

Platinum: virtually indestructible chemically at normal 90-percent platinum, 10-percent iridium, but still somewhat soft at Mohs $4^{1}/_{3}$. Ammonia turns solder joints dark. Iodine and mercurochrome both stain platinum.

Silver: at Mohs $2^{1}/_{2}$, easily dented or bent. Sterling is 92.5-percent silver, which tarnishes. Foams and liquids work but remove some silver and the dark outline that gives silver its character. Cleaning cloths are safer. Regular use and wiping limit the need for cleaners. Use no ammonia.

Miscellaneous Gems

A number of gems routinely used in jewelry are little known to the public. Some are rare and valuable but overshadowed by the famous, major gemstones. Several of the gems pictured here have become very big sellers as a result of TV shopping channels.

ALEXANDRITE (right), the color-change chrysoberyl, is typically bluish-green or yellowish-green in daylight and orangy-red or brownish-red in incandescent light. Usually ultrasonic cleaner- and steamer-safe, it is heat and light stable, with no chemical reactions. Hard at Mohs $8\frac{1}{2}$, it has excellent toughness. Clean with soapy water, alcohol, or cleaning solutions.

APATITE (left) is a surprisingly bright gem that comes in a color range from orange to purple. The pinks may fade in bright light. Because it is Mohs 5 with only fair toughness, do not chance ultrasonic cleaning. Never use a steamer or expose to acids. It is best to clean with only warm soapy water.

AZURITE, often confused with malachite and lapis, is a blue or violet mineral used for beads, cabochons, and carvings. Soft at Mohs $3\frac{1}{2}$-4, with poor toughness, it is heat- and acid-sensitive. Clean only with cool soapy water, never with ultrasonic cleaners or steamers.

CAT'S-EYE CHRYSOBERYL is the yellowish-green or brownish-yellow chatoyant variety of chrysoberyl. The moving white line that looks like a cat's eye accounts for its appeal. This is a considerably rarer and more expensive gem than the similarly-named but more plentiful cat's-eye quartz. Clean cat's-eye chrysoberyl the same as alexandrite.

CHROME DIOPSIDE is the green variety of diopside, a gem best known for its four-rayed black stars. Relatively soft at Mohs $5\frac{1}{2}$-6 with poor toughness, it is stable to light but sensitive to high heat. Clean chrome diopside in warm soapy water, not ultrasonic cleaners or steamers.

DANBURITE, usually colorless or pale yellow, has good toughness and light stability at a relatively hard Mohs 7. Since it is sensitive to high heat, danburite is best cleaned only with warm soapy water. Ultrasonic cleaners and steamers are risky.

HEMATITE, a smoky-black iron crystal that does not rust, is popular for affordable dark beads, cabochons, and cameos. Hematite, pyrite, and marcasite are the gem forms of iron. A durable material, with excellent toughness, hematite can be soaked or washed. At Mohs $5\frac{1}{2}$-$6\frac{1}{2}$, it is also safe in alcohol, ultrasonic cleaners, steamers, and most chemical cleaners.

IOLITE (left), an affordable blue or violet gem, is often mistaken for sapphire or tanzanite. With fair toughness and good hardness at Mohs 7-7½, it is often used in rings. Avoid acids, sudden temperature changes, ultrasonic cleaners, and steamers. Clean iolite with warm soapy water and a soft brush.

KUNZITE (right) is a pink to bluish-purple spodumene named for the great 19th century American gemologist, George Frederick Kunz. Available in large, clean stones, it is relatively inexpensive. Medium hard at Mohs 6½-7, it fades in sunlight. Cleavage planes make it vulnerable to blows. Clean only in warm soapy water with a soft brush.

MARCASITE & PYRITE are both iron sulfite. Pyrite is best known as "fool's gold," a collectible crystal mass of fused cubes. Smaller cubes are used in jewelry. Genuine marcasite turns brown over time in light, so today's jewelry marcasite is actually almost always pyrite. Both materials have good toughness but are relatively soft at Mohs 6-6½. They can be cleaned in ultrasonic cleaners or steamers or with warm soapy water.

MOLDAVITE & ST. HELENS GLASS are two forms of glass used in jewelry. Moldavite, a variety of tektite, is a natural glass believed to be created when a meteorite impacts earth and melts surrounding sand. The glass that results is then cut into gems. St. Helens glass is a man-made material said to be created by melting ash from the slopes of the volcanic Mt. St. Helens. Since both these materials are as breakable as regular glass and obsidian (natural volcanic glass), they should be treated gently. Glass has poor toughness and is soft at Mohs 5-6. Avoid high heat and all chemicals, some of which can etch surfaces or dissolve glass. Clean only in warm soapy water with a soft brush.

MOONSTONE can display adularescence (a light play beneath the surface), asterism (a 4-rayed star), or chatoyancy (cat's-eye light line). This feldspar should be handled carefully because of its poor toughness and Mohs 6-6½ softness. Clean only with warm soapy water and a soft brush.

RHODONITE, "the stone of love," is a pink or red mineral that is often found in very large boulders. It is plentiful enough to be used for tabletops, bookends, and vases. For jewelry, it is usually cut as cabochons or beads, but is soft at Mohs 5½-6½. Transparent, faceted gems are rare. Avoid high heat, rough handling, and acids. Some tightly compacted material can be safely cleaned in both ultrasonic cleaners and steamers. But, as always, the safest cleaning technique is warm soapy water with a water rinse.

ZIRCON was the favorite diamond substitute before the creation of CZs. Available in a wide variety of bright colors, these natural gems are often confused with other stones. Medium hard at Mohs 6-7½, with only fair to poor toughness, zircon's main appeal is its very high refractive index, which produces great sparkle. The safest cleaning method is warm soapy water.

Opal

Whether they originate in Australia (left) or in Oregon (above), opals are relatively fragile gems. To be safe, wipe them with a dry or damp cloth and store separately from other jewels.

Opals are among the best loved and least understood gemstones. Almost everyone has an instant attraction to the fantastic rainbow effects in the depths of an opal. The opals that get the most attention have great play-of-color, a phenomenon caused by light diffraction from thousands of tiny silica spheres. Some of the best and most expensive opals display prismatic colors dancing above a black or dark body color. More common and less expensive is white opal, material with white or light body color. Others are fire opal (sometimes called Mexican opal, in red, yellow, or orange, with or without play-of-color), and jelly or water opal (colorless or nearly transparent with little or no color play).

Opals are softer and more fragile than most crystalline gems. Be careful not to scratch or hit opals, especially those mounted in rings. Because opals are composed of three to twenty percent water, they should not be allowed to freeze or dry out. The shock of moving from a warm home to frigid winter air can crack them. Avoid plastic bags and dry storage conditions. Soft cloth bags with padding are ideal for opals. The water in opals should be encouraged to stay there by storing your gems in even, high humidity.

Clean opals either with a soft dry cloth or a damp moist one. Do not wet doublets or triplets or soak opals. Avoid all rough treatment and ammonia. With care, you can enjoy wearing opals for years.

Ultrasonic: do not use
Steamer: do not use
Warm soapy water OK
Hardness: 5 - 6½
Toughness: very poor to fair

Do not soak or bleach
Avoid chemicals
Fragile; avoid impacts
Avoid very dry conditions
Light: stable

Heat: opals contain some moisture. High heat and very dry conditions can cause cracking. Avoid sudden or extreme temperature changes

Pearls & Cameos

Pearls, the premier organic gems, are likely the oldest gems of all. Because they emerged from oceans and rivers ready to wear, virtually every primitive society collected and wore them. Until the beginning of this century, all pearls were naturals, the result of saltwater oysters, freshwater mussels, conch, and abalone responding to irritants. Unable to expel a piece of shell, coral, or other debris caught inside, the creature coats it with smooth mother-of-pearl, or nacre. Over years thousands of concentric layers of nacre form a natural pearl.

Since the early 1900s, virtually all pearls have been cultured. Freshwater pearls, most of which are still solid nacre, are cultured in mussels. Technicians place pieces of living mantle tissue into the mantle of a live mussel to stimulate pearl growth. To make saltwater pearls, shell beads are surgically placed inside oysters as an irritant and nucleus. Thus, today's cultured saltwater pearls and most round freshwater pearls are shell beads coated with nacre. The thickness of that coating determines luster and longevity. Mabes (mah-bays) are made by gluing plastic domes inside shells, which oysters then coat.

Mother-of-pearl (MOP) buttons, once common, now appear only on quality clothes. MOP beads for necklaces and earrings are cut mainly from larger South Sea oyster shells. Iridescent abalone shell is increasingly used in jewelry. Traditional cameos are carved from Bahamian conch shell.

Care for mabes, MOP, conch cameos, and abalone as you do for pearls. Acids attack pearls, which are alkaline. Avoid contact with hair spray, perfume, alcohol, cosmetics, bleach, ammonia, swimming pools, and even acidic perspiration. **Always** put on pearls *after* applying perfume, cosmetics, and spray. When stringing, knot between pearls to reduce abrasion. With heavy wear, restring annually, otherwise every two years.

Wipe your pearls after every wearing with a soft, dry or damp cloth. If you must wash your necklace, do not rub pearls together. Rinse, pat with a clean cloth, then stretch out the strand on a soft cloth or towel to dry.

Ultrasonic: do not use
Steamer: do not use
Mild soap with water OK
Hardness: $2^{1}/_{2}$ - $4^{1}/_{2}$
Toughness: usually good, but variable
Attacked by all acids
Avoid perfume and sprays
Heat: can burn or cause cracks
Separate from other jewelry
Light: stable, except when dyed
Other care: knot all good pearls; restring every two years; keep clean; wear pearls touching clothes instead of skin; keep away from chemicals

Peridot

Peridot (pronounced PEAR-ih-doe) is an affordable green gemstone that may have a much grander history that most owners realize. Mined today principally on American Indian reservations in Arizona, it is also found in a variety of countries, including Burma, Brazil, and Sri Lanka. A source in the Red Sea, an island called Zabargad (now part of Egypt), may have produced the world's first great green gems. References in the Bible and other ancient texts to large clear green jewels have been translated as "emeralds." But emeralds from Cleopatra's Mines in Egypt were neither large nor clear. Most likely the world's best green gems before Europe's introduction to Colombian emeralds in the 1500s were actually peridot.

Softer gems, like peridot, require extra care. Gemstone abrasion is usually caused by two factors. Some damage occurs when softer gems come into contact with harder jewels. Avoid scratches by storing gems and jewelry in separate cloth bags after each use. Other damage occurs, particularly in rings, when gems softer than Mohs 7 contact hard everyday objects.

Clean peridot with warm soapy water or with a few drops of mild detergent in water. Use a soft brush if needed. Rinse. Dry with a clean cloth. Avoid any sudden temperature changes, such as a warm house to frigid winter temperatures, and avoid contact with acids.

Ultrasonic: considered risky
Steamer: do not use
Warm soapy water OK
Hardness: 6½ - 7

Avoid acids
Abrasion possible in rings
Toughness: fair to good
Light: stable

Heat: uneven heat or sudden temperature changes may cause fracturing

Ruby & Sapphire

Rubies and sapphires are different colors of the same material. All the gems in the ultrasonic cleaner are Montana sapphires; the centerstone above is a ruby.

Rubies and sapphires are among the oldest gemstones people have used in jewelry. Although the public perception is that they are two separate gems always colored red and blue, rubies and sapphires actually are color variations of corundum, the crystal form of aluminum oxide. Sapphires come in every imaginable hue except one. When corundum crystals are red, caused by the presence of chromium, those crystals are rubies.

Rubies and sapphires may be the perfect gemstones. At Mohs 9, they are harder than all other gems except diamonds. The color range of corundum is complete: ruby's red and sapphire's blues of all shades, plus purple, yellow, orange, green, pink, brown, gold, gray, teal, black, and even colorless, which is increasingly used for accent stones. To add to their appeal, sapphires are tough, stable, and more affordable than diamonds.

Most damage to rubies and sapphires occurs from rubbing against diamonds, from hard knocks directly to the stones, and from facet and girdle abrasions caused by careless handling. Even though they are among the most durable of gems, abuse can cause some scratches and chips.

Cleaning these gems (either naturals or synthetics) is easy and straightforward. Because they are tough and hard, almost any cleaning technique is acceptable. You can normally use ultrasonic cleaners, steamers, soapy water, and brushes. Do not put heavily fractured gems into mechanical cleaners. Because rubies and sapphires are far more durable than gold, silver, or platinum, your main concern should be the safety of the metal.

Ultrasonic: safe except for oiled rubies*
Steamer: safe except for oiled rubies*
Warm soapy water OK
Toughness: excellent, except with fractured gems
Heat: usually no problem, because almost all have been heated already
Light: usually stable, except for irradiated orange and yellow sapphires

Chemicals seldom affect them
Hardness: 9
Unusually hard, stable material

*No mechanical cleaning for oiled rubies, heavily fractured stones, or black star sapphires

Spinel

Spinels are paradoxically among the most beautiful yet least recognized of all gemstones. Often mistaken for other gems, spinels deserve to be treasured for their own merits. Although found in Sri Lanka, Thailand, Cambodia, and Russia, the most valued specimens come from Burma. In addition to brilliant blues and reds, spinels occur in pink, orange, yellow, violet, and purple.

Even though 10th-century Arabic scientist Al-Birumi recognized spinels as separate gems, people continue to confuse rubies and red spinels. Old European cutters knew the difference because rubies are harder, yet red spinels were called "balas rubies." Many great red spinels in royal collections across the Continent are still mislabeled. By the time Henry V placed it on his battle helmet and went on to victory in 1415 at Agincourt, the Black Prince Ruby had already enjoyed fame in Spain and England. Its glory later caught the eye of King James, who had it set in the State Crown, after which it joined the famous Cullinan Diamond in the Imperial State Crown. Now on display in the Tower of London as part of England's history, the Black Prince Ruby is actually a 170-carat red spinel. Today, Switzerland manufactures most synthetic spinels, the majority of which are used in U.S. class rings.

For safety, separate spinels from diamonds, rubies, and sapphires, which are harder. It is the metal in spinel jewelry that has to be protected. Spinels are such fine gems that almost nothing will harm them.

Spinels are among the easiest of all gems to clean. Hard and tough, they do not react to chemicals. Therefore, mechanical cleaners, soapy water, alcohol, ammonia, and almost all household and jewelry cleaners work.

Ultrasonic: usually safe
Steamer: usually safe
Warm soapy water OK
Hardness: 8
Heat: intense heat can fade light-colored stones

No reaction to chemicals
Hard and durable gems
Toughness: good
Light: stable

Tanzanite

Tanzanite, named by Tiffany's for the gem's origin, is new to many jewelry buyers. Seldom does a beautiful and popular gemstone enter the world market. But in 1967 miners in northern Tanzania discovered a deep periwinkle-blue zoisite, which was practically an instant success. This new gem is strongly trichroic. As you view the jewel from different angles, the facets dance with electric hues from purple to blue.

Most heat treatment intensifies hues and/or clarifies internal inclusions. Even though it is possible to find examples of natural-colored tanzanite, the vast majority on the market is a result of heating greenish-brown zoisite. The treatment is permanent, leaving no visible indication that the material has been heated.

Tanzanite needs extra attention. Two characteristics affect its care. First, because it is softer than quartz, the usual precautions apply. Protect tanzanite rings from being scratched by other gems, metals, stones, rocks, and walls. Facet edges easily abrade. Second, tanzanite has internal cleavage planes that can part if struck. It is also somewhat brittle, like emeralds. Although perfect for pins, earrings, and pendants, tanzanite set in rings is vulnerable, so be gentle.

Wash your tanzanite only in warm soapy water, and do not use mechanical cleaners or chemicals. Dry with a soft cloth.

Ultrasonic: do not use
Steamer: do not use
Warm soapy water OK
Hardness: 6 - 7

Attacked by some acids
Toughness: fair to poor
Light: stable
Heat: fuses under very high heat

Topaz

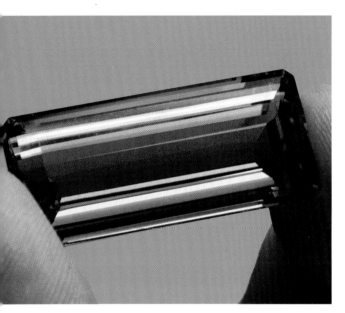

Topaz is available in a variety of colors and prices, from rare and relatively expensive imperial topaz (left) to one of the most popular and affordable gems, blue topaz (above).

Topaz is a family of beautiful gemstones with a very broad price range. Before the discovery in the 1980s that irradiation turns widely available and inexpensive colorless topaz into electric blue jewels, most of the market centered on collectors and colored stone enthusiasts. Especially popular were Brazil's reddish-orange imperial topaz and rare pinks from Pakistan. Now far more colorless (or white) stones are sold than all colors combined. Most are treated to produce a variety of blue hues, and recently, white topaz is often used as accent and even centerstones.

Whether the gem is a popular golden, an easily affordable blue, or a rare imperial or pink, care and cleaning are the same. The main concern is that topaz easily cleaves. It can split from even a light blow to a cleavage plane or from the pressure of being set too tightly. A sudden change of temperature may also cause stones to break. So, although topaz is hard at Mohs 8, it requires gentle handling. Give it the same kind of care you would give emeralds. Topaz is harder than other gems except diamonds, rubies, sapphires, and spinel, so store it separately to protect other jewelry.

The easiest way to clean topaz is in warm water with soap or a drop or two of mild detergent. Rub small, stubborn areas with a soft toothbrush. Rinse, wipe dry, then air dry on a towel. Handle topaz carefully.

Ultrasonic: do not use
Steamer: do not use
Warm soapy water OK
Hardness: 8
Heat: rapid cooling or heating can cause breakage; avoid hard knocks; easy cleavage makes mechanical cleaning and heating risky

Slight reaction to chemicals
Avoid rough handling
Toughness: poor
Light: stable

Tourmaline

Tourmalines offer yet another gem family surprise. Everyone knows they are green, but few buyers realize that tourmalines come in several other colors. In fact, tourmalines occur in virtually every color, including colorless, brown, yellow, and orange. Indicolite is the blue or violet variety. Red or pink is called rubellite. Multicolor or parti-color tourmaline shifts to two and sometimes three different colors in one crystal. Occasionally "watermelon" tourmaline crystals grow with a red interior and a green rind. Brazil is the best known source, but the gems are also found in Afghanistan, Burma, India, Sri Lanka, Africa, and the U.S.A.

Tourmalines are relatively easy to care for. Because they are only slightly harder than quartz and softer than topaz, spinel, rubies, sapphires, and diamonds, you should keep tourmalines in individual cloth bags so that other jewelry will not scratch them. Avoid high heat, which can alter colors, and sudden temperature changes, which can cause fractures. Those are both unusual, but possible, occurrences. Normally tourmalines are trouble-free.

The old reliable cleaner, warm soapy water, works fine with these gems. Because tourmalines do not react with chemicals, you can also use mild detergents or alcohol, which soften stubborn spots that can then be brushed away with a soft toothbrush. Finish cleaning with a water rinse and wipe dry.

Ultrasonic: considered risky
Steamer: considered risky
Warm soapy water OK
Hardness: 7 - 7½
No reaction to chemicals
Reds, pinks have more inclusions
Toughness: fair
Light: generally stable
Heat: high heat can alter colors; sudden changes can cause fractures
Multicolored stones may be weaker where colors join

Turquoise

Turquoise is often associated in the U.S. with Native American silver jewelry. Globally, the material is thought of as an expensive, high-quality Persian gem, which is often an intense warm medium blue without webs. Egypt produces a yellowish-green webbed turquoise. Far more commonly sold is U.S. or Mexican light blue, greenish-blue, and bluish-green turquoise, usually with spiderweb patterns. Because American Indian artists highly prize beautiful webbed U.S. material, the prices are steep.

Like all such soft, porous minerals, turquoise is a delicate material that requires some care. It is very sensitive to heat and chemicals. Avoid both. It can absorb anything that touches it—perspiration, cosmetics, perfume, skin oil, cleaning fluids, ammonia. Unfortunately, those chemicals turn turquoise green or a muddy gray that cannot be reversed. Most turquoise today is "stabilized," impregnated or coated with acrylic, lacquer, or epoxy to hold it together and to lessen the effects of chemicals. Be aware that because it is soft, other jewelry or contact with harder objects may abrade turquoise.

Metal cleaners that are safe for silver will discolor turquoise. To clean turquoise, wipe with a soft damp cloth and dry immediately. Do not soak. Keep it away from all cleaning fluids. Use gentle care only.

Ultrasonic: do not use
Steamer: do not use
Warm soapy water OK
Heat: avoid all heat; use great care when setting stones
Chemicals: avoid all chemicals; even perspiration and cosmetics are risky

Light: usually stable
Toughness: fair to good
Hardness: 5 - 6

Mohs Hardness Scale

Named for German mineralogist Friedrich Mohs (1773-1839), the Mohs scale is widely used as a hardness standard for gems and minerals. The concept is simple, employing what is often called the "scratch test." Once Mohs determined that diamonds were the hardest of all materials, he assigned that gem the number 10. He then sought the softest mineral he could measure, talc, and gave it the number 1. Mohs rubbed hundreds of materials together. The one that scratched the other received a higher number on his scale. By repeating the test with new samples, Mohs established a relative rating for all the common gems and minerals. Knowing this concept emphasizes the importance of keeping harder gemstones from scratching softer gems and jewelry metals.

Mineral	Hardness	Mineral	Hardness
Talc	1	Marcasite	6-6½
Soapstone (steatite)	1-2½	Benitoite	6-6½
Alabaster	2	Pyrite	6-6½
Amber	2-2½	Nephrite	6-6½
Serpentine	2-4	Tanzanite (zoisite)	6-7
Silver	2½	Kunzite (spodumene)	6-7
Gold	2½-3	Zircon	6½-7½
Jet	2½-4	Andradite garnet	6½-7
Pearl	2½-4½	Jadeite	6½-7
Coral	3-4	Peridot	6½-7
Calcite	3	Chalcedony	6½-7
Conch pearl	3½	Danburite	7
Azurite	3½-4	Quartz	7
Malachite	3½-4	Grossularite garnet	7
Rhodochrosite	3½-4½	Iolite	7-7½
Platinum	4⅓	Andalusite	7-7½
Fluorite	4	Tourmaline	7-7½
Smithsonite	5	Spessartite garnet	7-7½
Apatite	5	Pyrope garnet	7-7½
Serpentine (bowenite)	5-5½	Rhodolite garnet	7-7½
Obsidian	5-5½	Almandite garnet	7½
Sodalite	5-6	Emerald (beryl)	7½-8
Turquoise	5-6	Topaz	8
Lapis lazuli	5-6	Spinel	8
Glass	5-6	Cubic zirconia	8½
Diopside	5-6	Chrysoberyl	8½
Opal	5½	Ruby (corundum)	9
Moldavite	5½	Sapphire (corundum)	9
Rhodonite	5½-6½	Silicon carbide	9¼
Hematite	5½-6½	Diamond	10

About Fred Ward and his Gem Book Series

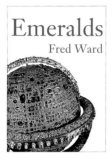

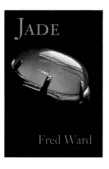

Glamour, intrigue, romance, the quest for treasure... those are all vital aspects of humankind's eternal search and love for gemstones. As long as people have roamed the world, they have placed extraordinary value on our incredible gifts from the land and sea. And once they found or purchased gems, owners discovered their treasures lasted longer and were more beautiful with proper care. The purpose of this book is to offer simple, straightforward, accurate information on the safest and easiest methods for wisely taking care of your gems and jewelry.

Gem Care is the fifth in a series of gem books written and photographed by Fred Ward. Each book, *Rubies & Sapphires*, *Emeralds*, *Diamonds*, *Pearls*, *Gem Care*, and *Jade*, is part of a 16-year global search into the history, geology, lore, and sources of these priceless treasures. He personally visited the sites and artifacts pictured in all the books to provide the most authentic and timely information available in the field. Fred Ward's original articles on gemstones first appeared in *National Geographic* Magazine. In addition to being a journalist, Mr. Ward is a Graduate Gemologist (GIA), the highest academic achievement in the gem trade.

Mr. Ward, a respected authority on gems and gemology, is in great demand as a speaker to professional and private groups. After years viewing the gem trade around the world, he formed Blue Planet Gems with designer Carol Tutera to make his vast experience available to others. Blue Planet Gems specializes in fine custom-designed jewelry and private gem searches. In addition, he formed Gem Book Publishers to print and distribute these books. And, because of his long interest in jade, he founded Jade Designs to create and market a line of nephrite tabletops, bookends, and other large hand-crafted jade pieces.

For those interested in printing mechanics, this book is part of the on-going computer revolution of desktop publishing. It was designed entirely using PageMaker 5.0a electronic layouts on a Mac Quadra 950. *Pearls* was printed by H & D Graphics in Hialeah, Florida, using Adobe Janson typefaces.